FRANCIS O. J. SMITH'S REPLY

TO

THE ARGUMENT OF HON. AMOS KENDALL,

IN THE

MATTER OF ARBITRATION

OF

MORSE & AL. AND SMITH.

T. R. WALKER, and H. O. ALDEN, Esqrs., Arbitrators.

THOMAS R. WALKER, AND H. O. ALDEN, Esqrs.,
Arbitrators, &c.

GENTLEMEN:

The elaborate printed argument of Mr. Kendall in this matter, occupying 32 pages, with a manuscript addenda of several pages, undeniably proves to the minds that arbitrate, *one*, or *both*, of *two* positions:—

1*st*, That the defence set up against my claim, is of a very complex and occult nature—a sum of so many elements that a very large slate is required to work out its results: *or*,

2*ndly*, It is a very difficult and doubtful defence, requiring a great variety of experimental views and arguments, to secure any degree of success to him who makes it.

But, to my understanding of the subject, the facts are neither numerous nor complicated. The deductions to which they point are simple, direct, and easy of comprehension. Indeed, the space and labor of a correct statement of them are small, compared with the space and labor which are required in following and refuting the fallacies and inconsistencies with which Mr. Kendall has surrounded them, in his argument and manuscript *addenda.* These, it may be deemed expedient that I should follow and refute ; without imitating, however, his attempts at ridicule, or sarcasm, as they can furnish no rule of judicial decision to arbitrators.

In this connection, it may properly be remarked, as a fact made evident in the proofs before you, that, from the time Mr. Kendall made his contract with me, in *April* 1847, for an exchange of portions of my New Orleans Line interest, for Morse and Vail's Boston Line interest, carrying with it a power from me, to receive the Stocks representing the former, to the full amount of the exchange made, he never did, at any time, *down to the date of this arbitration*, render to me any account whatever in detail, of the Stocks he had so received on my account, nor of the amount which he claimed, or was entitled so to receive !

Equally certain is it, that from the time he made a further contract of exchange with me, for a further interest in my W. and N. O. Line Stocks, for Morse and Vail's Buffalo and Milwaukee interest, he never rendered to me any specific account of what he had received, or of what he claimed to receive, under such further contract.

The whole evidence of the solitary piece of information which he had given me on either subject, *at any time*, is to be found in the fact, that in the suit of Morse and Vail *vs.* myself, he had credited to me, upon the Milwaukee exchange contract, the receipt of $21,239.30 ! He now says that was erroneous, and it should have been $22,239.30. *See his pamphlet, p.* 14.

How much he had previously received for the Boston Line, or how he had arrived at that credit of $21,239.30 for the Milwaukee Line, no where was ever made known to me; nor is it now pretended, that any account of Stocks received by him, under either contract, was ever rendered to me by him, *during the seven years pendency of the suit above named!*

The first glimmer of an explanation on either subject is to be found in his letter to me of *January* 1859, which will be found on pages 13 and 14 of his printed argument.

It is in evidence on p. 7 of that pamphlet, that as early as June 1848, I complained to him of my ignorance of his manner of administering upon my Stocks in the W. and N. O. Line.

On page 7 of the same pamphlet, it is in evidence, that in his reply to my complaint, he was not prepared to give me any intelligible data on the subject.

Even as late as *November* 1858, almost ten years later, in a letter of his to me, which I have already submitted to you, it is made apparent, that I was "*seeking knowledge under difficulties*" upon these *obfuscated* subjects, with only the promise of light and information *at some future period.*

On the 28*th of Dec.* 1858, with the best lights which I could obtain up to that date, I submitted to Mr. Kendall a statement, from which he has made an extract on pages 31 and 32 of his argument; and which statement, I informed him in a subsequent letter, contained some "radical errors," and that there were also some in his letter to me of the *5th of Nov.* 1858.* Without producing this subsequent letter, or his own of the *5th of Nov.*, wherein errors were avowed to exist, he is enabled to indulge, on page 16 of his argument, his proclivity to sarcasm, in comparing the *erroneous* data with the *corrected ones of a subsequent date;* and as if the intervening discovery of error, so kept out of view, could not

* See extract from this letter, onward.

be summoned up, to render his sarcasm alike unfounded and ridiculous.

But, to illustrate Mr. Kendall's own disposition to make his claim "grow amazingly within the last year", I need only refer—1st, to his letter to me of Jan. 1859, on pages 13 and 14 of his argument, wherein he concedes a sum of $5,491.28, which had not been, but still was to be credited on the Lake Line, if he could get that judgment re-opened—being not previously accounted for—while, 2dly, he *now* claims to sink all that, and render me indebted, in his manuscript addenda, (p. 19), $9,148.60—a modest growth of $14,639.88 in his account since January, 1859!

In him, who had all the keeping of this stock, such floundering from correct results would seem to be inexcusable; while on my part, I have ever been kept without any definite knowledge, though for ten years and upwards I had been entitled to and seeking it.

In his letter of *Jan.* 1859, quoted on page 13 of his argument, is displayed a reference to a *stock account* therein enclosed, as if it could furnish any information upon the disputed stock account, upon which I was seeking of him information. I submit *that* document in manuscript herewith, marked *A*, to show you that it *explains nothing respecting the stocks which had been received by him for Morse and Vail*, under either of my before named contracts with them, and so conveys no information at all.

On page 17 of Mr. Kendall's argument, he expresses *surprise*, that I should quote from his letters of *May and June* 1847, and of *Jan.* 1859, for rules to govern the arbitrators in this case. And to enforce his expression of this surprise, he there adds:—"Perhaps Mr. Smith can explain the appplica-"bility of these quotations to the points in issue, or their utility "for any purpose; but I am unable to perceive the one, or "appreciate the other!"

But, if the Arbitrators will compare *the* Mr. Kendall who thus exhibits himself on *page* 17, with *the* Mr. Kendall who exhibits himself on *pages* 6 *and* 7 of the same pamphlet, they will find that the Mr. Kendall on pages 6 and 7, not only deemed the said letters of May and June 1847 as applicable to the subject matter here in controversy, without any emotion of surprise ; but, after *re-quoting* them in detail, adds the following emphatic commentary thereon, conceding to them *the same importance which I ascribe to them in this case*, viz :—

" Here is a special contract expressly excluding these lines " from the exchange contract, and providing that the stock " accruing thereon, should be issued one fourth to Mr. Smith, " and three fourths to Messrs. Morse and Vail. *It is only " under this contract that Mr. Smith can now claim any stock " upon those lines.*"

Consistency is an admitted jewel, whether looked for in *principle*, or in *argument.* But to establish its existence on comparison of the pages of Mr. Kendall's pamphlet referred to, another pamphlet of at least an equal number of pages *will be found necessary.*

I quoted these letters to refute the idea, that the exchange contract contemplated a conveyance to M. & V. of $\frac{645}{1172}$ parts of my interest in the *entire* 1716 miles of the W. and N. O. Line, as subsequently constructed.

Mr. Kendall, in quoting them, concedes, that the elongation of the line to *Charleston* and *Savannah* would not be construed as embraced by my conveyance, *and so constituted none of the* 1172 *parts of the Line really conveyed.*

But, I quoted them for the further purpose of showing, in the language of Mr. Kendall's letter to me of *June* 27, 1847, that not only the elongation to Charleston and Savannah were excluded from the exchange, but that all other elongations "*of the same character*", whereby the Line had been extended beyond the 1172 miles originally agreed on as the basis of our exchange, were excluded likewise.

If a departure from the original Post Office route in one case, justly excluded such elongation from the 1172 parts originally made the basis of exchange, can any reason be assigned, why *every other departure* involving an elongation from the 1172 miles *should not also be excluded*, particularly if the Patentees' stock was thereby proportionably increased?

Mr. Kendall attempts to find a reason for such an inconsistency, by stating for a fact what is no where proved to be a fact—viz: "Neither line *was, or ever expected to be*, confined "to the *mail routes;* and as the W. and N. O. Line was not "built or located, the post office distances presented the best "criterion then available for measuring the relative value of "the two lines" *K.'s pamphlet, p.* 4.

Now, at the time the exchange contract was entered into, the post office route was the route that had been built on between Boston and New York; its longest line, to this day, is not twenty miles variation from it. And hence the exchange was based upon *an equal distance* of the post office route between Washington and New Orleans. It was an exchange of stocks on the basis of *a mile of one for a mile of the other*. Hence, all the aggregate miles of the W. and N. O. Line, upon the *mile for mile* basis, over the amount requisite to pay M. and V. for their aggregate of miles in the B. and N. Y. Line, (viz. 645 miles of 1172, the whole distance contemplated,) *was agreed to be unaffected by the exchange contract*, and so defined. Morse and Vail's aggregate of miles in the B. and N. York Line being equal to 161 and 1-4—or three-fourths of 215 miles—and calling for four times that number of miles in the W. and N. O. Line to furnish from my interest the equivalent of 161 1-4 miles, or a total of 645—all over that 645 of the W. and N. O. Line *was agreed to be* *unaffected by my contract of exchange.* The agreed length of the *W. and N. O. Line*, at that time, being that of the post office route—"assumed" to be 1172 miles, the excess to be so unaffected by the exchange was consequently stated at 527 miles.*

In other words—it was only 645 miles out of the total of 1172, that was thus agreed to be made the absolute property of M. and V. by our contract of exchange. In all the excess, my one-fourth interest remained unsold. And why should it not be so? Stock to the Patentees was awarded for every such elongation. The stock being based upon the length of the line—and the post office route being, alone, contemplated, *at that time*, for that length—the *mile for mile* basis of exchange was very properly defined by 645 out of the 1172, the assumed route.

That I am right in claiming that the exchange contract was understood and intended by both parties to be based upon an *equality of miles* on which stock should be issued, I need only advert to the admission which Mr. Kendall has forced himself to make, in his argument, at page 4, where he says—

"For the purpose of this exchange, *it was assumed*, that "the value [of the two lines] *was the same*, MILE FOR MILE, "adopting the post office distances as the basis of the as- "sumption."

Suppose it had been known, at the time the exchange contract was made, that the New Orleans Line would have gone by Charleston and Savannah and back, thus increasing the distance 284 miles as the basis of stock; can the Referees doubt, upon Mr. Kendall's before quoted admission of the exchange being made upon the "*mile for mile*" basis, that the ratio of my conveyance to M. and V. would have been defined as $\frac{645}{1456}$ parts of the line, instead of $\frac{645}{1172}$ parts?

---

* This assumed distance was an error of singular magnitude for Mr. K. to fall into, or to lead me into. See letter of Ass't P. M. General, onward.

All doubt as to this must be considered as removed by the actual proceedings of the parties, to meet the emergency of a departure from the post office route, as soon as it arose. No contemplation of such a departure is to be found in the original contract of exchange. No provision was consequently there made, for a basis of division among the patentees, in case of such a departure. And yet, as soon as any departures were suggested, as to Charleston and Savannah, it was at once *conceded by Mr. Kendall, and agreed,* that the increased distance was not covered by the exchange contract. In all stock to represent such increase, I was to retain my undiminished quarter ownership, and the rule of $\frac{645}{1172}$ parts was to be applied only to the *original route.*

Nor, in so settling this meaning, were the divergences from the P. O. route, or main line, to Charleston and to Savannah *alone* embraced; but, this settlement was *to admit all "others of the same character", so as not to affect the principles of the exchange contract.* See Kendall's letter to Smith, *May* 27, 1847, *K.'s argument, p.* 6.

Can, or *need* any evidence be more conclusive, as to the original meaning of the exchange contract?

If so, have we not even got it, in Mr. Kendall's pamphlet, p. 4, where his language is—"For the purpose of this exchange, it was assumed, that *the value was the same,* MILE "FOR MILE, adopting the post office distances as the basis of "the assumption."

M. and V., on the mile for mile exchange, were to receive the whole stock, or interest in 645 miles of the W. and N. O. Line—which was assumed to be 1172 miles long—making it $\frac{645}{1172}$ parts of the whole.

But, when that *assumed* length of the W. and N. O. Line was found to require an elongation, the elongated parts were at once agreed not to belong to the 1172 parts embraced by the exchange.

Still more strikingly illustrative of this meaning of the parties is the fact, that when Mr. Kendall discovered, that the stockholders had not only elongated the line 142 miles, to Charleston and Savannah—but had DOUBLED the elongation, by issuing stock to both contractors and patentees, for a line going to *both* those places, and *for a line coming back from both places*, he still CONCEDED, that *no portion of this duplicated elongation stock was covered by my exchange conveyance to Morse and Vail.*

[*See his concession of these points clearly made on page* 9 *of his pamphlet.*]

I may here repeat, with emphasis, the inquiry—" If a departure from the original post office route in *one* case—or in *two* cases—justly excluded such elongation in each case, and *even excluded a duplication of them*, from the 1172 miles originally made the basis of exchange, can any reason be assigned, *why every other departure*, involving an elongation from the 1172 miles, *should not also be excluded*, PARTICULARLY IF THE PATENTEES' STOCK WAS THEREBY *proportionably increased?*"

It is not enough to say, that such departures *diluted* the value, by increasing the aggregate amount, of the line's stock. For it was not so. The *quid pro quo* is found in the acquisition to the line of business points, not within the original 1172 miles, and of value equal to such increase of stock. The original territory only was designed to be assigned, and not any enlarged territory, nor stock to represent such enlarged territory.

How can I be said to have assigned my interest in territory which did not at the time fall within the contemplation of the line.

The whole idea must be set down as an *after-thought*, to which the insatiable and constantly growing and absorbing cupidity of Mr. Kendall must stand godfather.

To maintain, that a contract of conveyance embraces *that*, which, *at its date*, was not, *and had never been thought of by either party* to the contract, and so was not provided for as a contingent interest—is as absurd to the logical, as it is untenable to the legal, and unjust to the equitable mind.

But there are certain positions for the guidance of the arbitrators, upon which the adverse parties are indisputably agreed.

They are—

1st. That "the issue of Stock to Messrs. Morse and Vail and Kendall on account of exchange of interests, shall be adjusted upon the basis of *miles*, as provided in the original contract." *See Mr. K.'s argument, page* 3.

Mr. Kendall re-states the same position on the same page, as follows: "The Arbitrators are instructed to adjust the Stock between the parties growing out of the exchange arrangements, just as if it had never been issued."

2nd. "For the purpose of this exchange it was assumed, that *the value was the same*, MILE FOR MILE, adopting the Post Office distances as the basis of the assumption." *See K.'s argument, page* 4.

3d. "The Post Office distance from New York to Boston was 215 miles, and that from Washington to New Orleans was 1172 miles." *Ibid, page* 4.

4th. "As Morse and Vail's interest in both lines was three times that of Mr. Smith, it would take three times as many, or 645 miles of the Washington and New Orleans Line to pay for their interest in the 215 miles of the Boston Line." *Ibid, page* 4.

5th. "The Lines from Branchville to Charleston, and Millen to Savannah, not being portions of 'the main line of communication,' do not form a portion of the 645 parts of the Line on which Mr. Smith conveyed his entire interest to Messrs. Morse and Vail." "The question was settled by a specific agreement that these Lines should not be considered a portion of the main line embraced in the exchange arrangement." *Ibid, pages* 6 *and* 7 *and Letters there quoted.*

6th. The reported length of the Line from Washington to New Orleans was 1716 miles, at the time of the organization of the Company. "But this distance includes the side Lines to Charleston and Savannah *twice over.* The wire was run down to those points from Branchville and Millen, and back again on the same posts, and the length of wire was counted in as so many miles of Line," by the Stockholders in settling with the Contractor, *and in the issue of Stock to the Patentees. Ibid, page* 6.

7th. The distance so duplicated is 141 miles, making an aggregate of 282 miles in the 1716 miles, leaving for the main line 1434 miles as the length of the main line, instead of 1172 miles if it had followed the Post Office route. *Ibid, page* 9.

8th. These 282 miles "should be deducted from the 1716 miles, leaving 1434 miles as the length of the main line from Washington to New Orleans." *Ibid, page* 9.

9th. All Stock issued on account of the Patentees, *on both the first and second wires,* between Washington and New Orleans, has been issued to Morse and Vail in their own right, to the extent of three-fourths, and to my account to the extent of one-fourth—[less $ 7,500.] *Ibid, pages* 10 *and* 11.

10th. The dividends on the entire Patent interest have been, in like manner, received exclusively by Morse and Vail. *Ibid, page* 7.

11th. The total of Patentees' Stock issued on first and second wires, was as follows—

On first wire from Washington to Petersburg,	$21,450
Ditto from Petersburg to N. O.	235,950
On second wire to Petersburg,	7,150
On ditto beyond Petersburg,	41,731
Total,	$306,281

12th. Only $21,239,30 of my W. and N. O. stock was credited or accounted for to me by M. and V. on the exchange contract for the Milwaukee Line. *Ibid, page* 14.

I believe, Gentlemen, there is nothing in the preceding twelve propositions, controverted by the parties before you.

The dispute then arises from, and is limited to, the claim set up by Mr. Kendall, to draw within the operation of my conveyance to Messrs. Morse and Vail, in exchange for their interest in the 215 miles of the Boston Line, not only all my interest in three times that number, (or in 645) miles, in the W. and N. O. line—(although that is proved, and conceded to be the meaning of the contract, in propositions 2 and 4, stated above from Mr. Kendall's argument—) but, 645 parts, (or 789 miles) of the whole 1434 miles of the line, remaining after deducting the duplicated 141 to Charleston and Savannah, are claimed to belong to M. and V. under my exchange contract for the Boston line.

Mr. K. thus words his claim:—

" Dividing the line *according to the contract*, it appears, " that the whole patent right on 789 miles of it belongs to " Messrs. Morse and Vail, who obtained Smith's interest in " exchange for their interest in the New York and Boston " Line, leaving 645 miles on which Mr. Smith retained his " one-fourth." *See his pamphlet, p.* 39, *at bottom.*

Taking these asserted data of Mr. Kendall, as the basis of your decision, and the unavoidable result is simply a controversy between the parties, *whether the main line of the*

W. and N. O. route, which is to be the basis of an exchange of interest by them, ought to be considered 1434 *miles, all above being out of dispute—or* 1172 *miles, and putting the difference between these last sums, (or,* 262 *miles) also out of dispute?*

If the latter be the correct decision as I maintain it is, then 645 parts of 1172 miles (or that proportion of the stock) became the ***absolute*** property of M. and V., and Smith was left one-fourth owner in 789 miles of the 1434.

If the former be the correct decision, as Mr. Kendall maintains it is, then 645 parts, or 789 of 1434 miles (or that proportion of the stock) became the absolute property of M. and V., and Smith was left one-fourth owner in 645 miles only, of the 1434.

I have already stated all that can be needful to establish the correctness, the justice, and original admission by the parties acting in concurrence, of the construction and rule of the exchange contract, for which I still contend.

Mr. Kendall has referred to no one concurrent act of the parties at any time imputing any different construction or meaning to that contract.

In no one instance does Mr. Kendall show that he has acted consistently with the rule he now contends for. He has been acting under an erroneous appropriation of my stock in the W. and N. O. Line from the beginning, and so never could render me a specific account of his doings. To give color to these misappropriations, he is obliged to set up a succession of equally groundless and false theories of accounts.

Let me illustrate his positions, and disprove even the latest and most improved edition of his bad stewardship, in several particulars.

Illustration I. In his pamphlet, pp. 13 and 14—by way of accounting for his alleged credit to me of $21,239.31 (made without my knowledge) on the Milwaukee Line—he admits that he appropriated the whole of my one quarter of the *whole* first issue of $264,550, except $7,500—viz: $\frac{645}{1172}$ parts of $66,137.50—being $36,398.19, in payment for the Boston Line; and $21,239.31 more—(but says it should have been $22,239.31)—to the Milwaukee Line. These two sums, added to $7,500 issued to myself, in part for $10,000 specially reserved, make a total of $66,137.50—the quarter of $264,550.

To make the last sum, 1716 miles are covered by stock at $150 per mile for 1st wire—amounting to	$257,400
And for stock on 2d wire between Washington and Petersburgh,	7,150
	$264,550

See Mr. K.'s Argument, Doc. F, *p.* 28.

But observe, this includes *my one quarter of stock on the* 282 *miles to Charleston and Savannah,* conceded by Mr. Kendall *not to have been conveyed by me,* nor subject to his appropriation! Has he explained away an error now made so palpable? That amount, or 1-4 of 150 per mile on 282 miles, thus *freely* appropriated against the law of our contract, and without my knowledge, is $10,575.

Illustration II. On page 10, you will find Mr. Kendall's amusing process of bringing me in debt to the amount of $5,551.80 of stock, alleged to be *over-paid* to me, when all the evidence on the Company's books, and all other evidence in the case, shows, that the $7,500 of stock, (it should have been $10,000) expressly reserved for issue to me in the contract E, page 26 of Mr. K.'s argument, is *every dollar of patentees' stock ever issued to me on the W. and N. O. Line!*

Mr. Kendall's *peculiar* process of getting at such a result is, 1*st*, by omitting from the amount and keeping out of sight, the excess of $10,575, which he had without authority appropriated as above illustrated, of the stock conceded to belong to me outside of the Boston line exchange contract—

and 2*dly*, by omitting from the account and keeping out of sight *all credit to me for the stock he had received of the first issue for the second wire, from Washington to Petersburg!*

To add to the amusing character of this exhibit, by Mr. Kendall, of my indebtedness, he proceeds on the three next succeeding pages of his argument, to get up a theory by which, *although not a dollar had ever been issued to me on account of the second wire beyond Petersburg*, and M. and V. had received all of that part of the patentees' stock, I am brought in debt in another sum of $2,257.87 over-paid me!

If any system of "*growing claims*" can partake more of the absurd and ridiculous, Barnum, of humbug celebrity, ought to have the man who can invent it, for a show in his museum!

But no more need be added on my part, than to state the accounts according to each theory of construction of the contract, and leave the impartiality of the arbitrators to adopt that which is the just one, to their understanding.

1st. If the contract of exchange was originally understood and framed by the parties to be an exchange of interests in the two lines, "*mile for mile*"—or, as of a like value in the ratio of their length for the issue of stock—as *propositions* 2 *and* 4 above deduced from Mr. Kendall's own argument verify to be the true construction—then Messrs. M. and V. are entitled, under the exchange contract, to only $\frac{645}{1172}$ parts of the stock issued on 1172 miles—and Mr. Smith's ownership continued absolute for one-fourth of the excess, being 544 miles.

Messrs. Morse and Vail's account with him, therefore, must stand thus:—

Dr. To an aggregate receipt of stock on Mr. Smith's account—(not including $7,500 issued to himself)—on 1st and 2d wires—being one-fourth of $306,281; or 76,570, less $7,500, - - $69,070.00

Amount of Debit, brought forward, $69,070.00

Cr. By due on Boston Line exchange, 645 parts of stock issued on Smith's 1-4 of 1172 miles; computed thus—The whole sum of $306,281, divided upon 1716 miles, makes an average of $178.48 1-2 per mile, and upon 645 miles amounts to $115,122.82; 1-4 of which, assigned by Smith, is - - - - $28,780.20

By credit in suit on Milwaukee Line, - 21,239.31

Total of credits, 49,939.51

Balance in Morse and Vail's hands, not accounted for heretofore to Smith, - - - $19,130.49

To which is to be added dividends received, and interest on same to date of award.

2d. If the contract of exchange was originally understood and framed, without any controlling reference to the respective values of the lines based on distances, but, 282 miles of divergence to Charleston and Savannah were nevertheless specially excluded from the exchange, as Mr. Kendall concedes, leaving the line 1434 miles long, instead of 1172, and of which my one-fourth in 645 parts thereof were absolutely conveyed to M. and V. for the Boston line, then their account with me would stand thus:—

Dr. As above, - - - - - - $69,070.00

Cr. By due on Boston Line exchange, 645 parts of stock issued on Smith's 1-4 of 1434 miles; computed thus—The whole sum of $306,281, divided upon 1716 miles, makes an average of $178.48 1-2, and upon 1434 miles amounts to 255,947.49; one-fourth of which is $63,986.87, one 1172d part of which sum is $54.59, and 645–1172d parts amount to - - - $35,210.55

being the credit for Boston Line payment.

By credit in suit on Milwaukee Line, as before, 21,239.31

Makes a total of credit, 56,449,86

Deducted from debit, leaves in Morse and Vail's hands, due Smith, - $12,620.14

To which is to be added dividends received and interest on same to date of award.

This last result is deduced under the rule laid down by Mr. Kendall on p. 9 of his argument, in which he gives me 1-4 of 645 miles, besides the Savannah and Charleston quota of stock on 282, and M. and V. 789 miles, in this way :—

Debit—645 miles, at $178.48 1-2 per mile, being the ratio of the whole stock issued on 1716 miles, for first and second wire, would amount to $115,122.87; 1-4 of which amounts to		$28,780.71
282 miles to Charleston and Savannah, at $178.48 1-2, amounts to $50,333.51; of which 1-4 amounts to		12,583.37
Making a total of stock, exclusive of 789 appropriated wholly to M. and V.—received by them,		$41,364.08
Credit—Against which is only chargeable to me—		
Stock issued to me,	$ 7,500.00	
And stock credited to me on Milwaukee Line,	21,239.31	
Total of Credit,		28,739.31
Balance due Smith,		$12,624.77
Result of first process,		12,620.14
Difference from computing fractions in last process,		$ 4.63

One question only remains in respect to the preceding results, viz. a question of *interest*. The rule Mr. Kendall, in his written *addenda* to his argument, under date of Jan. 13, 1860, claims to be justly applicable to the settlement of these accounts is the rule that ought to be satisfactory. It is, to make the party who has received the dividends of the other party account in interest on the same, at the rate of six per cent. per annum, from the time of each receipt. If the balance is in my favor, he cannot of course object to the rule he has prescribed for application, on the supposition that the balance was against me.

The table of dividends and interest, upon the basis of my

being entitled to recover the $19,130.49, will be as follows, to May 1, 1860 :—

Per ct.	Date of Dividends.	Stock.	Time.	Dividends.	Interest.	Total.
3	July 3, 1850	$19,130.49	9 y'rs 10 mo.	$573.91	$338.59	$912.50
3	July 5, 1851	"	8 " 10 "	573.91	304.16	878.07
3	July 20, 1852	"	7 " 9 "	573.91	266.84	840.75
5	July 20, 1853	"	6 " 9 "	956.52	387.47	1343.99
3	Jan'y 4, 1854	"	5 " 10 "	573.91	200.86	774.77
3	Aug't 1, 1854	"	5 " 9 "	573.91	197.99	771.90
2	July 2, 1856	"	4 " 10 "	382.60	111.45	494.05
2	Feb'y 4, 1857	"	3 " 2 "	382.60	71.15	453.75
2	Aug't 2, 1857	"	2 " 9 "	382.60	63.12	445.72
2	Jan. 26, 1858	"	2 " 3 "	382.60	51.63	434.23
2	July 12, 1858	"	1 " 10 "	382.60	42.08	424.68
3	Jan'y 2, 1859	"	1 " 4 "	573.91	45.83	619.74
3	July 11, 1859	"	9½"	573.91	27.25	601.16
3	Jan'y 1, 1860	"	4 "	573.91	11.47	585.38
				$7,460.80	$2,119.89	$9,580.69

Amount of stock, upon the "mile for mile" basis, $19,130.49
Dividends and Interest on same,* - - 9,580.69

Total due, - $28,711.18

* By the articles of submission, the stock found due is to be returned in stock, and the dividends and interest are to be refunded in money.

Gentlemen:

Since preparing the preceding reply to Mr. Kendall's argument, I have received, in answer to inquiries made on the subject, from the 2d Assistant P. M. General of the U. States, information in the subjoined letter, which places beyond any doubt the justness of the full claim which I have advanced upon Messrs. Morse and Vail for the one quarter interest in all the stock issued upon all the line to N. Orleans, beyond 645, originally assigned them. It is for only precisely what they have received and retained upon the exchange contract for the Boston Line, beyond what that contract called for, and beyond all the credits that have been allowed me in the adjustment of the Milwaukee exchange contract.

Two points are now established in regard to the meaning of the contract of exchange.

1st. That the 282 miles, to Charleston and Savannah, were not embraced in the basis of exchange.

2d. That it was the *intention* and *understanding* of the parties that in all the distance of the line upon *the Post Office route, beyond* 645 *miles*, my one-fourth interest should be unaffected by the exchange.

The Post Office route was assumed, *because understood* to be, 1172 miles only. That turns out to have been *an entirely erroneous and false assumption.* Mr. Kendall represented the distance, and so, I doubt not, understood the P. O. route to be. *But it was not so in fact.* Both parties are clearly proved to have been misinformed. It was 1407 miles—and hence the line extended to 1434 miles, by including the diversions to offices in different towns. It was the intention of *both parties*, as Mr. Kendall must admit, to base the exchange upon the *truth* of the case. If it involved an error, " clear and unquestionable," *will he insist, if he might, upon advantage from such error?* He says, p. 8 of his argument, " The arbitrators are now to revert back to the first issue of stock in 1848, and adjust it, so far as regards the exchange of interests, *precisely as they would then have adjusted it.*"

Can there be any doubt, that in such adjustment, on discovering the *assumed* distance of the post office route to have been wholly erroneous, the correct one would have been *then* adopted as the basis of apportioning the stock under the contract?

If Mr. K. should insist otherwise, the arbitrators can be bound by no such wrong demand. All that Mr. Kendall in fact has pretended to claim is, that 645 parts of the length of the post office route between W. and N. O., as it was in 1847, are covered by the intentions of the parties to the exchange contract, and " mile for mile" in payment for their interest in the Boston Line.

This would seem to end all controversy. The 282 miles conceded for Charleston and Savannah, and 1407 now ascertained to be the correct post office route, makes an aggregate of 1689 of the 1716 miles—difference 25 miles only—a small per centage in 1716 miles for the *zig-zag* course of the line, (all of which is included in settling with the contractor, and allowed accordingly in the issue of stock to the patentees,) upon the highways, and varying from the centre line of the highways.

Believing the amount so delineated above, being of principal $19,130.49, and of dividends and interest on dividends $9,580.69, making a total of $28,711.18, thus incontestibly established as my just claim, I subjoin the Assistant P. M.

General's letter referred to, as conclusive upon all doubts—viz :—

POST OFFICE DEPARTMENT,
CONTRACT OFFICE, April 13, 1860.

SIR :

The distance by the post office records from Washington by [70] Fredricksburg, [60] Richmond, [24] Petersburg, [64] Weldon, [18] Enfield, [18] Rocky Mount, [41] Goldsboro', [29] Warsaw, [56] Wilmington, [178] Charleston, [64] Branchville, [75] Augusta, [78] Union Point, [28] Madison, [26] Covington, [42] Atlanta, [40] Newnon, [32] Lagrange, [15] West Point, [28] Auburn, [60] Montgomery, [46] Greenville, [46] Burnt Corn, [71] Stockton, [22] Blakely, and [12] Mobile, to [164] New Orleans, is 1407 miles, according to the course of the mail in 1847.

Very respectfully, E. L. CHILDS,
For Second Ass't P. M. General.

F. O. J. SMITH, Esq.,
New York.

But, since the issues of all stock of which I had knowledge, there seems, by the letters of Mr. Kendall to and from Mr. Geo. Wood, herewith submitted in copies marked *C*, *D*, *E*, *F*, that there has been made, as claimed by Mr. Kendall, on my account, additional issues of stock, *not hitherto accounted for to me*, amounting to - - $4,024.37

Total, - 5,932.00

$9,956.37

(*See letters C, D, and E, in MSS.*)

Of the above, three-fourths, as Mr. Kendall claims in said letters, belonged to Morse and Vail in their own right, leaving $2,489.09—of which sum $\frac{645}{1716}$ parts belong to Messrs. M. and V. on account of Boston Line exchange—being $935.25

Leaving the excess due me, amounting to - * 1,553.84

* This amount is payable in stock, according to the articles of submission.

On which the *average* interest in lieu of dividends is due, as upon the previous indebtedness—or say five years, - - - - - *$456.16

Of Mr. Kendall's letter *F*, I have only to remark, in reply to his hypothetical comparison of what Messrs. M. and V. gave, with what they have received, *for their Boston stock;* had he only copied *all*, instead of *part* of my letter to him of Dec. 28, 1858, (see his argument, p. 31,) his entire hypothesis would have been refuted, without any additional word from me. I subjoin the part of *that* letter which he *cautiously omitted* to print, viz:—

" The *whole* of Morse and Vail's Boston stock on the two original wires from New York to Boston, was 401 shares—representing a like amount of cash shares at $50 each paid in—amounting to $20,050; *but issued to represent* $75 *each*, and amounting to $30,075.

" For this, on the two first wires of New Orleans stock, you will have received, as above, $36,397.35
and 5,741.69

$42,139.04

representing that sum paid *in cash* by subscribers—(being $20,050,) making $22,089 in New Orleans stock *more than all the Boston stock* derived by me from M. and V. for their interest in the Boston Line, or more than two dollars for one!

" In view of this, I do not think you will deem any part of my claim, as above stated, in the N. O. Line, as unjust."

In my letter to Mr. Kendall of Jan. 1859, I said:—

" In my letter to you of 28th ult. there were some radical errors—as also in yours to me of the 5th of Nov. last. *I did not have our Boston Line contract before me*, nor your letters, for my guidance; and it impressed me strangely, that the 1716 miles of New Orleans line should produce me so little beyond paying for your three-fourths of 215 miles of the Boston line. I think I have now the data to exhibit a correct result." &c.

---

* This is payable in cash, as provided in the articles of submission, making a total of $2,010.00.

Abating the error which the December letter to Mr. Kendall, written under the want of correct details from Mr. Kendall, as before explained on p. 3 of this reply, and still, the actual exchange of my N. Orleans stock, reduced to the cash value as paid in by subscribers, for the M. and V. stock in the Boston line, also reduced to cash value as paid in by subscribers, actually gives Messrs. M. and V. a very large excess in the W. and N. O. Stock over what they give me in the B. stock, if all the exchange be settled precisely on the basis I now claim. And while I ask nothing that is not clearly right—nothing in the shape of penalty, of forfeiture, or of hardship, beyond the hardship of a surrender of that which has been received by, but does not rightly belong to, the other party—I am confident your careful and just discrimination will not require me, under the finaliter of this arbitration, to submit to any thing that is wrong.

All of which is most respectfully submitted.

FRANCIS O. J. SMITH.

April 30, 1860.

I. Berry and Son, Printers,
177 Fore Street, Portland.

www.ingramcontent.com/pod-product-compliance
Lightning Source LLC
LaVergne TN
LVHW011144110826
845150LV00008B/2506

* 9 7 8 1 4 1 8 1 9 2 4 6 4 *